Writing Down the Words

A Collection of Columns
by Pauline Clarke
Printed in the Berkshire Record
from 1990-1999

For additional copies contact:
Pauline Clarke
PO Box 374
Great Barrington, MA 01230

ISBN: 0-9677826-0-0
Library of Congress Catalog Card Number: 99-98124

Printed in the United States by:
Morris Publishing
3212 East Highway 30
Kearney, NE 68847
1-800-650-7888

Forward

The following articles were written between 1990 and 1999 and were published in the Berkshire Record, a weekly paper in Great Barrington, MA.

Several readers have told me they're tired of collecting newspaper clippings and asked wouldn't I please put my columns into a book of some sort. This is it.

I couldn't fit them all in this attempt, so there's apt to be a Volume II at some point. This first collection is dedicated to Lora Atherton with gratitude and love.

Collection Contents

Really Good Stories

includes:

•A Really Good Story
•Up a Tree
•The Great Road Game Walk
•Spikey's Home
•Dashing Through the Snow
•Pork Chops Linoleum and Other Dinner Disasters
•The Truth About Dinner

A Really Good Story

I've always loved a good story. If it's really good, I'm not adverse to hearing it over and over, which probably explains my penchant for telling the same tale repeatedly. My children (and later Bob) have been long-suffering but I can always tell when I've overdone it. The corners of their mouths begin to twitch, their eyes glaze over, and in a voice that could win Oscars one of them will say, "We know, Mom. We *know*."

It's not surprising then that when, year after year, my children dressed for their prom dates, I wistfully recalled that I had never been invited to a prom. When child number three came down the stairs modeling her dress a few days before her big date, I sighed.

"I never got to go to a prom..."

"We know, Mom. We *know*," came the weary chorus.

Bob looked at them and then at me. "Did I miss something?"

"You're lucky. You only heard it once," said Jennifer, twirling so a cloud of white tulle flared out around her legs. "Mom says that *every* time one of us goes to a prom."

Bob sought to console me with a man's point of view. "Proms are usually overrated. The anticipation is probably the best part."

I remembered seeing a picture of Bob and his date posing at their senior prom in front of a gargantuan flower arrangement. Her gloved hand rested possessively on his arm. The smile on his face was more than expectation called for.

"Well," I said morosely, "I didn't even get to anticipate."

What happens next is a tale worth retelling. Bob came home a few nights later and said to me, "Start anticipating."

I gave him a blank look. He grinned. "Cissy, would you like to go to the prom with me?"

"You mean as a chaperone?" I asked, thinking of Jennifer's prom. "I'm sure that's all been arranged."

"Not as a chaperone, as my date," he said.

"We can't go to Jennifer's prom!"

"No, no," he laughed. "The Summer Singers theater group is sponsoring a prom to raise money for props. It's open to the public. You can dress up if you want. I bought two tickets."

I was beside myself. I'd just been asked to a prom.

Jennifer floated off to her dance on Friday evening and on Saturday I dressed for my own. I pulled a borrowed dress of blue taffeta over my head, took the pin curls out of my hair, brushed it into a wavy cloud around my face, borrowed my daughter's eye makeup, and slipped my feet into a pair of satin dancing slippers.

Bob whistled when he saw me. From behind his back he pulled a corsage. I ran to the refrigerator and took out the boutonniere I'd hidden there. He looked splendid in his black suit and white silk tie. I felt like Cinderella.

He'd been right about the anticipation—the food was poor, the music was awful, and we were almost the only ones dressed up, but I'd never had a more wonderful evening. On the way home I thanked him for what must have been the hundredth time. He gave my arm a squeeze and said with a grin, "Now you can't tell that story to your kids anymore."

"No," I replied, "but this one's even better."

Up a Tree

I came home the other day to find Bob up a tree.

"What are you doing up there?" I asked.

"Trimming branches," he replied. Just then a gust of wind made the tree sway. Bob clamped his arms around the trunk. "This is just like an amusement park ride," he called down as the tree bent first one way and then the other.

"Yikes," I muttered, watching him. "The man is crazy." I went into the house.

Sometime later I went back outside to ask Bob what he wanted for dinner. I looked for him in the garage and in the garden. I circled the house before I thought to look up. There he was, still in the tree. He was trying to hook the top rung of the ladder with his foot. It had slipped from its position and leaned crookedly against the adjoining tree.

"What are you trying to do?" I asked him as the ladder extended rapidly and threatened to come apart.

"The wind moved the ladder. I'm trying to bring it back where I need it so I can get down. The darn thing will come apart if I pull it too hard."

"Do you have another ladder? I could use it to climb up on the roof and reposition yours."

He thought a minute. "No," he said, finally. "But if I can drop this one down could you climb up and throw me a rope?"

I don't usually climb ladders. No, that's not strictly true. I can climb ladders okay. It's getting off them (or back on them) that I don't do. But, Bob was stuck. I kept that in mind while I fetched the rope and watched him move the ladder about with his foot until it slid off the edge of the roof. I guided it to the ground and bravely started to climb.

When I reached the top rung, I looked up at Bob. *What do I do now?* I wondered, knowing that I would never be able to get off the ladder onto the roof.

"Get off the ladder onto the roof," Bob said.

"I can't," I replied. "My legs won't move."

"Then climb down and see if Cassie can come out and help," he suggested.

Gratefully, I climbed down and ran into the house to get Cassie. She came outside, looked up and shook her head. "What are you doing up there?" she asked Bob.

"Trying to get down. Can you climb up the ladder and throw that rope to me?"

"Sure," said Cassie and she climbed quickly up to the top rung. She looked up. "What do I do now?" she asked.

"Get off the ladder and on to the roof," Bob said.

"I can't," said Cassie. "My legs won't move."

All three of us were quiet for a moment, thinking. At last Bob said, "Could you girls extend the ladder all the way and move it so it's leaning against my tree?"

"Sure," we chorused, glad not to have to climb the thing again.

Cassie clambered down and grabbed one side. I grabbed the other. "I'll push, you pull," I said. "We'll move out and then over."

Together we stood the ladder straight. I looked up. The rungs reached 26 feet into the air. I looked at Cassie. "Ready?" she asked.

"Ready," I agreed but neither of us was prepared for the weight of the ladder. It swooped toward the ground. We raised it back up only to have it swoop in the other direction.

"We can steady it or we can move it but we can't do both," gasped Cassie as we danced about the lawn, the ladder between us. Eventually we danced close enough to the tree to rest the ends of the ladder against it. The thing stretched out almost horizontally to the ground.

"We have to hike it up some," I pointed out. Bob swayed back and forth in his tree, watching us struggle.

"You two will never be asked to join 911," he remarked from his perch.

We pushed and jimmied and shoved until the ladder reached as far up the tree as it could. There was still a good twenty feet of tree trunk between Bob and the top rung. Cassie and I watched breathlessly as he clasped the trunk in both arms and began to shimmy down a few inches at a time. Each time he slipped we gasped. When his foot finally touched the ladder we both cheered, "Hooray!"

Back on the ground, Bob dusted off his hands and said, "I felt like a kitten stuck up there."

He looked a little worse for wear. There were long scratches on his arms and pine needles in his hair. He had several puncture wounds in his chest and only later decided some of his ribs must have cracked as he hugged the tree coming down.

I've heard of being out on a limb or up a creek without a paddle or what goes up must come down. Only Bob would be up a tree without a ladder.

The Great Road Game Walk

There's nothing like a springtime walk in the country. Ah, the fresh air, the warm sun lying like a benevolent arm across your shoulders, the breeze on your face—it's a great way to spend time. I walk both for the exercise and the peacefulness. A good deal of my writing is done in my head as I walk. I stride along at a fairly good clip, pausing now and then to take in the view or duck under a fence to detour through a meadow. Most often I walk alone and my favorite jaunt is the one I take early on Sunday mornings. I leave Bob snoozing in bed and step out into a world that's hushed and quiet. Only an occasional car passes. Church bells toll in the distance, and it seems I have the world to myself. If I time my walk right, I catch the birds at feeding time. On the unsettled stretch of road between here and Hamilton's Corner, the roadside is lined with old crabapple trees. There in early spring the birds gather to feast on last year's fruit and bees bumble and butt in a froth of apple blossoms.

There are times, though, when Bob wants to come along. Then a walk in the country takes on a whole new meaning. It becomes the Great Road Game Walk. (Note: Great Game Time is not limited to walks. Great Game Time occurs anytime Bob is around.)

We start out sedately enough. I tame my pace because if I trot along at my regular gait Bob complains that he can't *see* anything. And seeing is one of the elements of the Game. The rules go like this. If you see something truly interesting, like a stick ("What can I do with this?") or a piece of wire ("Can I use this for something?") or a feather ("Isn't this *neat* ?") you pick it up.

Long, sturdy sticks are the article of choice because they are all-purpose objects. You can play Kick the Stick, a game that depends heavily on timing and accuracy. Hold the stick in front of your right foot and without breaking stride, kick the end of the stick again and again and again, making an annoying whock, whock, whock sound as you walk. Or how about Drag the Stick? It's easy, really. You just drop the end of the stick in the gravel at the edge of the road and drag it noisily along behind you, leaving a wiggly line that you then make up a story about. ("Cissy! See that line? What could have made it?") There's Whack Things With Your Stick. You can whack anything—rocks, cans, leaves, other sticks, people. When you're tired of playing with your stick you can play Hide The Stick so no one else will find it and destroy it before your next walk.

Kick the rock is another Road Game favorite. One person can play but it's more fun if there's someone else along. The object is to kick the rock ahead of you in a catty-corner motion so that the other person can kick it back to you, all the while never missing a step. Cries of "NO FAIR!" can be heard in the next county if you take more than three kicks to get the rock back into position for the other fellow. This game can go on indefinitely or until one person gives the rock a mighty kick into the heavy undergrowth.

We never come home from walks together without some useful object like a stick decorated with worm trails, or a stone shaped like a pyramid or a feather whose flying days are over. Every moment has been filled with noise and games. If you're ever bored, come for a walk with us. I'll teach you how to play Break the Stick.

Spikey's Home

I have a little yaller (that's the Vermont pronunciation of yellow) dog who is, like the Green Mountain State he comes from, hardy and indomitable. He is also very old. If he has crammed seven years of dog life into every human year he's been alive that makes him almost 98. A bit long in the tooth to go gallivanting.

But that's what he did last weekend, probably figuring that he may as well live until he dies. Or maybe he was just following his nose and got lost. The poor fellow can't see more than a few feet in front of him and is as deaf as a post.

He still loves his food, though, which is how I came to notice he was missing in the first place. He usually comes nosing around the front porch early in the morning looking for a cookie and a bowl of water. If I'm not quick about it, he totters off to the stream out back and gets himself a drink. I catch a glimpse of him as he passes the window where I sit reading. I'm expected to have a dog biscuit ready when he returns.

When he didn't show up out front on Sunday morning, I went looking for him. I checked his house first to see if he was sleeping in. The weather had taken a cold, nasty turn and I slept later than usual myself. His blanket lay in a scruffy heap atop a flattened pile of hay but he was not on it. Nor was he in the pen—a bit of fenced off yard near his house.

I followed his well worn path from his house to the stream, fearing that perhaps his old legs had finally given out and that I would find him collapsed at the water's edge. He was not there.

Bob joined in the search and together we looked in all the places we thought he might be. Bob even got in the car and drove slowly up the road looking for signs of him. There was no point in calling his name. We'd only yell ourselves hoarse and he wouldn't hear us anyway.

By noon my anxiety was tinged with irritation. "The old coot," I muttered as I took off my sodden coat and wet boots. "Probably smelled some young pooch in heat and has gone off to make a fool of himself."

The next day, when he hadn't returned, my irritation turned to dread. He was, after all, my puppy, my companion for the past fourteen years, my confidant and my mainstay. When no one else liked me, his tail still wagged at my homecoming. When I didn't like anyone else, he was a major exception. When we walked together through the woods and fields, he never scampered too far ahead or failed to return to my side for a pat before dashing off after some new scent.

Monday morning I called the Dog Warden. He was out but late in the afternoon he called to say he'd been notified by the police on Saturday to pick up what looked like a dead dog on the side of the main road in town. When he bent to look at the dog, he was greeted by a growl.

"Not dead," he told me, "just tuckered out. I put him in the van and took him to the vet's."

I called the vet. "We're just closing," she told me. "But tell me what your dog looks like and I'll go see if it's him."

So, I described Spike. A yellow, curly coat, white front legs, an endearing Benjy face and ears that stood out like airplane wings when he was interested in something.

"He would never willingly get in a car," I said, remembering we had to drug him to get him here when I moved from Vermont. "And he doesn't have a collar on. He puts up a tremendous fight. He acts like I'm strangling him. Maybe it isn't my dog after all."

She advised me to come to the animal hospital early the next morning and take a look. Bob followed me in the van, thinking I might need help getting him in the car if the dog should be Spike. We waited while a girl was sent to the pens to fetch the stray. Moments later, she was hauled into the room by a very excited yaller dog at the end of a taut leash.

"Spikey!" Bob and I chorused. He didn't hear us but he smelled us and his joy knew no bounds. Faster than I could whip out my checkbook, he was out the door with Bob. I could see them together, Bob with the leash and the dog wrapped around his legs. I dashed off a check for room and board and ran out to the car.

I slid under the wheel as Bob opened the passenger door. His plan was to lift Spike onto the back seat and close the door before he could turn around and hop back out. Spike had plans of his own. He wedged himself between the bucket seats, one leg draped possessively over mine, his head on my lap, his other paw on the gear shift. "Take me home," his brown eyes pleaded.

I was so happy to see him, I talked love talk to him all the way home instead of giving him the scolding I'd intended. I know he's bound to go sooner or later but for a while longer, I'll have my little yaller friend to keep me company.

Dashing Through the Snow

To see me standing still on a pair of cross-country skis, poles held jauntily at my sides, you'd think I was born to the sport. To see me in motion is to believe something else altogether.

I was introduced to cross-country skiing when I lived in northern Vermont. I owned a pair of converted Army rescue skis left over from maneuvers in the Alps during WW II; at least that was the story sold along with the skis. They were long, wide, flat boards, narrowed and turned up slightly at the toes. Shoe clamps had been screwed to the middle of each one. Not for me the sleek, slim, waxless numbers that I bought for the children. I needed stability.

No one who knows me would accuse me of being athletic. Too long of leg and too short of body to have the balance that marks an athlete, I tend to overcompensate with a lot of arm waving and yelling as I capitulate to the pull of gravity. When I was first learning to ski, I capitulated regularly. I fell down going straight ahead on flat ground. I fell going down a hill—and going up a hill. I fell while negotiating corners. Sometimes I fell down just standing still.

Undaunted, I kept at it. I didn't have much choice. I had four young children on skis who needed righting when their overstuffed, snowsuited bodies toppled over. As a parent who owned skis, I was recruited for school field trips to cross-country meccas. Vermont weather cooperated by providing ten months of winter (and two months of darned poor skiing). I skied in self-defense. Then I moved south.

For the first few years here, winter was a disappointing season. It was cold without the buffer of snow. Ice often coated the roads. I couldn't go outside and play; it was too cold to just stand around and there wasn't any snow for snowmen or games of fox-and-geese or sledding. My skis hung uselessly from the rack in the garage.

This year is different. This year it has been snowing almost every other day. The inches have added up to more than a foot. And this morning a vast, unmarked, expanse of white beckoned from the other side of the window.

I stood poised on my skis, my poles attached to my wrists by leather straps. I was on the back stoop—a level, stone adjunct to the rock wall that borders the driveway. In order to access the field, I had to ski through the back yard. But first I had to scale a snow bank.

Conventional skiing wisdom suggests standing sideways to the bank, placing one ski firmly on the top and hoisting the other leg up using the poles for balance. I am nothing if not conventional. I lifted one ski-clad foot and planted it on the top of the bank. Immediately, my foot began to slide forward. My lower foot began to slide backward. I executed a fancy little flip right there on the stoop and landed in a tangle of poles and skis. Muttering imprecations under my breath, I righted myself, disengaged my skis and shoes, and *climbed* up the snow bank.

Once out in the field, with skis and poles, arms and legs working in concert, I felt my confidence surge. This was fun. This was great! This—as I lay suddenly and inexplicably on my back—was a mistake. A car horn tooted from the road. No doubt the driver had seen my little trick. I waved a pole from my prone position.

I was determined to ski to the top of West's hill and descend along the slope at an angle, the better to control my speed. It would be a straight, downhill run. No trees, no rocks, no obstacle lay in my path. I could then loop back to my original trail and return home. The wind picked up a bit as I slid along. As the slope steepened, I began to stomp my skis, toes turned outward. From the back, I must have resembled a large duck in winter gear.

I crested the hill, and stood for a moment, poised and ready, just in case anyone was looking. I dug my poles in, bent forward a little at the knees, and pushed. I moved maybe an inch. I thrust my poles out in front of me, heaved myself forward, and suddenly I was sailing down the slope, my scarf flapping behind me in the wind. I picked up speed. *Gosh,* I thought to myself, *I'm going awfully fast.* My knees stiffened. I began to lose my balance. Arms waving, poles flailing, I did a scarecrow dance in the middle of the hill and traversed the remainder on my derriere, my skis facing helplessly skyward.

Tomorrow promises to be clear and sunny. Tomorrow I will have to go to work. Tomorrow my skis will hang uselessly on the rack in the garage. What a pity.

Pork Chops Linoleum And Other Dinner Disasters

Dinner time is usually pretty tame at our house. One of us cooks, all of us eat, and we take turns cleaning up. Occasionally, a magnificent effort will produce oooohs of appreciation, but on the whole, mealtime is pleasant and uneventful.

There are times, however, when things get out of hand, and dinner gets (literally) turned upside down. Take Saturday night, for example. My sister, Jeanne, came for supper. She brought a roasted chicken and some fresh fruit for a compote. While I scrubbed and wedged some potatoes, she spread a rich crumble over the sliced fruit, and the two dishes went into the oven to bake.

Daughter Cassie was setting the table, and I was in the yard when we heard a shriek from the kitchen. There was a great crash and clattering followed by dead silence. Then came the moan.

Oh, oh.

I stood in the kitchen doorway, surveying the wreckage. Jeanne was ankle deep in fruit slices and potato wedges. Purple juice splattered the floor, the cupboards, the refrigerator, the walls. The three of us looked at each other, horrified.

"I was pulling the oven rack out," explained Jeanne, "when, whoosh! both pans came flying out of the oven like greased pigs. I couldn't catch either one. It's like they planned it—you know, one dish said to the other, 'when she opens that door, make a break for it.'"

This was greeted with more silence. Then one of us giggled. Giggles led to guffaws. Finally, we were all hooting as we tried to salvage what we could of the food.

"You scoop," said Jeanne, handing me a spatula, "and I'll spoon."

As an afterthought she added, "I hope your floor was clean."

"Well, no" I admitted. "But it will be when we're done here."

This was not the first time I've had to scoop dinner off the floor. We spent the next hour picking grit off the potatoes and telling tales of other dinner disasters.

"Remember the porch-chops?" Cassie asked, recalling the time I turned a whole pan of pork chops upside down on the porch floor. What with the dirt and the dog hair, they were rendered inedible. The kids grumbled at the grilled cheese sandwiches I substituted. The dogs were more appreciative. Porch-chops were an immediate hit with them.

That reminded me of another pork chop peccadillo. There were ten hungry guests gathered around the table, sniffing happily at the aroma of pork chop pizziole. We heard the oven door open. There was a scraping sound, a grunt, and a sudden tremendous, squishy thud. Pork chops, tomato sauce, and cheese hung from the doorknobs, slid down the walls, and made a thick, gooey river on the kitchen floor.

Ah," remarked one guest, immediately sizing up the situation. "Pork-chops linoleum for dinner tonight."

Pork chops are not the only thing to take flight at mealtime. Several years ago we were living in the cellar while building our log cabin over our heads. My mother and sister joined us for Thanksgiving dinner.

Things were cramped in the basement. Two adults, four children, and thirteen dogs (eleven of them puppies) crowded into whatever space was left around the furniture and packing boxes. Add two guests and a twenty pound turkey and something is bound to happen.

Dinner was nearly ready. The table was set, the potatoes mashed, the cranberries chilled. The kids and their dad were out gathering wood to keep the cookstove fire burning. I lifted the heavy roasting pan from the oven, and as I did, the bird took flight. It landed at my feet in a great splash of grease and slid with astonishing speed directly toward the open mouths of eleven startled pups.

Mama dove for the bird, dumped it in the sink, and began pumping water furiously over it. Out of the corner of her mouth she said, "Don't say a word and no one will know."

Until today, no one has.

The Truth About Dinner

Pete sat looking at his dinner plate.

"It looks like one of those gourmet meals in a fancy restaurant," he observed, giving me a slightly bemused look. "You know, the big plate with the small portions masquerading as enough."

I looked at my own dinner. Two slender hotdogs swam in an oniony sea of red chili sauce, a mound of crispy potato pieces sitting demurely to one side. A wide circular frame of white china surrounded the meager offering.

He reached over and touched my hand. "Thank you for making dinner," he said and I started to giggle.

The meal had been dredged up from a childhood memory triggered by the sight of the hotdogs in the refrigerator and the bottle of chili sauce in the supermarket when I dashed in for a jug of milk. I had a sudden yen for the supper dish my mother used to make that included lima beans in the puddle of red sauce, so I grabbed a box of those, too. Pete was adamant about no limas ("Nothing from Peru, thank you,") in his.

My mouth was watering by the time I sat down but the first bite sent childhood right back where it belonged.

"This is OK," I ventured, the unspoken "but" hanging in the air. Pete took another bite.

"You liked this as a child?" he asked, chewing slowly. "What do you call this dish?"

"A mistake," I had to admit, thinking memory is a chancy thing at best. "I liked it a lot better when I was ten."

There were other dinners I recalled looking forward to as a girl. One of them, a concoction of noodles, ham, and corn, I used to serve my own kids when they were young. When they suddenly blossomed into teenagers with voracious appetites and a hectic school schedule, I resurrected the make-it-quick casserole and surprised them with it one evening.

"What is this?" Bren wanted to know. He had a forkful of dangling noodles halfway to his mouth. Kernels of corn fell and bounced all around his plate.

"It's good, isn't it?" I asked. "You used to love this as a little kid," and I scooped some into my mouth.

We looked at each other over empty forks.

"I *liked* this?" he asked incredulously at the same time I said, "You *liked* this?"

The other kids got up to raid the refrigerator.

When you count up all the meals I've made over the last thirty years or so, two or three goof-ups don't constitute a bad record. Still, when dinner is not all your mouth and empty stomach hoped for, something seems to go out of the day.

"I was going to make stirred hamburger with milk gravy," I said to Pete as I shoved bits of my hot dog around in the chili sauce. The look he gave me stifled any further revelations. He got up and took the plates into the kitchen. I watched as he popped the few remaining potato pieces from the pan into his mouth. Even his back looked hungry.

Tomorrow I will thaw some chops, bake two sweet potatoes, make a salad. I will make dessert. I won't even mention that I have a yen for warm baked beans mixed with cold cottage cheese.

Reflections
includes:

- Gray Geese, Gray Days
- The Melancholy Season
- Caught Between Sunrise and Moonset
- Storms of the Heart
- Cathedrals and Grass Angels
- Signs to Live By
- WOW! Moments
- The Dreaming Bench
- Rain Reflections
- Rainbows and Imagination
- Life's Lessons

Gray Geese, Gray Days

I live not far from a lake. In early December the water is the color of pewter and cold; the ripples on its surface look like shivers. Sometimes the sun dances there but it does not bring warmth, only a shimmering illusion.

Great flocks of geese seek out the lake. With folded wings against rounded bodies, their heads dipped down beneath the surface of the gray water to feed, they look like floating rocks on a sheet of metal. I sit on the deserted beach and watch them. The cold, the geese, the dim lowering skies all speak of solitude and silence and the relentless approach of winter.

One day I watched the geese descend, their ragged, raucous vees coming apart as they splashed down, their wings outspread, their feet extended to break the plunge. In a moment they went from airborne to earthbound, their whole demeanor metamorphosed; wings folded and tucked, they were not so much bird as buoy. They gabbled quietly as they floated. Now and then a single goose would stretch its neck to the sky and flap its ponderous wings, flinging bright flashes of silvery water into the air.

The geese fly over the house often, their strange and haunting cries echoing over the rooftop. I stop what I am doing and look up, yearning for I don't know what, wishing, wanting...something. My response to their cry is visceral, ancestral.

Warmth and sunlight have flown with the geese. The shortened days begin in gray and end in gray. Early in the morning, before the reluctant sun opens its pale, distant eye, tree branches, bereft of leaf and color, stand in stark relief against a powdered sky. I watch as first light creeps into the yard, slowly spilling down the length of the great spruce that border the road, reaching with pale narrow fingers for the gate latch, the flagstone steps, the shingled wall. Mornings that the sun does not appear at all, the light merely flattens and spreads until the landscape is made visible. Spectre shapes become bushes and clumps of dead goldenrod anchored in the garden of spent summer dreams.

In its quiescent shades, the sky looks benign. The cold is almost a surprise beneath all that cottony cover, as if reaching into mittens you're fingers met ice cubes. Some days a chill wind blows, harbinger of the storms waiting to shriek down from the north. Other days the cold feels damp, as though with any encouragement the leaden skies would crack open and spill an icy rain.

I like the gray of early December. I like the silence and the cold and the muted, faded colors. The days are like pearls strung on silver thread, each one rounded and yielding to the shadow of the next. I listen to the geese and I yearn but not to leave, not to fly. I only want the moment to stay, to resist for a while the steady, insistent pull of the great seasonal wheel.

The Melancholy Season

Honor winter's lesson...long for the sun on your shoulders but let the frost and the cold come.

from *Journey to the Heart* by Melody Beattie

This is the melancholy time of year. The flowers are still beautiful but it is a fading beauty—a last burst of dazzle before their season is ended. The bright green leaves of summer have grown limp and tired and those few that flaunt their fall colors only serve to remind me that summer is dying. The sunshine, still warm at mid-day, leaves the sky too soon and there's a chill to evening's breath.

I feel like a child that has been sent to bed too early. *Wait!* I want to plead, just a few more hours. I haven't finished with the day yet. My pleading falls on deaf ears. The seasons will turn despite my fervent wishes. It is I who am out of step but how do I stop the ache at such beauty? It is all wrapped up in the delight, an intrinsic part of the joy. It is as though I am more acutely aware of the fact that though the beauty is real, the forms are not, and I want them to be. Because I have always resisted change, I find I must suffer it.

I've been this way for as long as I can remember. I was never ready for the season to change. I loved each day with a passion, and when darkness fell, I was reluctant to let go. Even as a small child, I knew the seasons' signals—the sudden cool breeze in

August that heralded November, the unexpected thaw in January that promised April. Anything I loved I wanted never to end. What I am beginning reluctantly to understand is that longing never keeps things close. It only lessens the immediate pleasure.

So, today I leave the apples heaped in the sink, the paring knife unused on the counter, and go for a walk in the fields. The sky is clear, cerulean blue, with feathery sweeps of cloud that look like abandoned angel's wings. I flop on my back in the soft, tawny grass and stare straight up. On either side of the sun stands a rainbow—sun dogs, guarding the light. I could sleep here in the meadow with the sun on my face and the crickets chirping all around.

Instead, I walk on, wending my way through the rows of field corn still uncut; tall, green soldiers forming a seemingly impenetrable phalanx. I find a narrow opening where the deer have broken through, and sneak in. I sit among the stalks where the air is still and warm, and hidden from view, I listen as a small breeze rustles the outer leaves. It can't reach me, deep in the sunlit greenery but I know what it is saying. It is whispering good-bye.

A week from now, the corn will be gone, leaving stubbled rows and a few flattened stalks that still bear brown-tasseled ears. The crows, the south-bound geese, the foraging turkeys will all come here to pick at the cobs. I will walk here, too, caught up in the beauty of the waning season, watching the colors of fall blaze and fade into buff and brown. In the midst of mourning summer, I will feel the fierce joy of winter's birth.

Caught Between Sunrise and Moonset

I wakened in a moonbeam slanting through the still dark sky though I could tell by the quality of gray that it must be after six a.m. The light came through the western window and lay in a contoured path over the quilt. I stretched out my hand and let the moonbeam lay in my palm; to have closed my fingers over it would have extinguished its magic.

I reached, instead, for my robe and slipped down the stairs to light the kettle for tea. With mug in hand, I went out into the early morning to watch the moon descend. Over my shoulder, the eastern sky was just beginning to pale. I stood entranced, looking first in one direction, then the other, never able to see both the rising sun and the setting moon at once, though I knew the waxing and waning lights were simultaneous and single sourced.

It put me in mind of beginnings and endings—how one is always the source of the other. As I turned full circle in the chill dawn, watching the growing daylight diminish the moon's brilliance, I could see between sunrise and moonset the workings of the world; always, always, the slow turning from beginning to end to beginning, on and on, the circle spiraling up and up without pause, always different, always the same.

My feet were touching ground they had danced on in childhood but the grass I stood on now was not the grass I danced on then—it had died and renewed itself a thousand times over. The last of the pink phlox that bloomed bravely in the waning season of the border garden were descendants of those my mother planted, no longer connected to her hands save through my memory. Caught between daylight and darkness, I was made to understand that the two exist simultaneously— it is only my limited vision that forces me to perceive one at a time.

The soft rush of traffic in the distance gradually became a muted thunder. Two cars passed me standing there, one headed toward the moon, the other toward the sun. A few small birds twittered in the roadside hedge, crows called across the meadow, and a small, cold breeze brushed my face. I kept turning in awe, watching the light change the landscape into everyday ordinariness, realizing suddenly that it had been extraordinary all along.

Try as I might, I could only face one direction. The maples that fringed the feet of the western mountains glowed scarlet in the fading moonlight. The pines that protected the brook to the east lifted their spires into a rose-colored sky. For a moment the entire world was washed in gold-tinged pink. The moon and sun faced one another, one haloed in blue-gray mist, the other surrounded by pale blue light. Caught in the middle, I turned around slowly, face lit first by moon glow and then by sunlight until at last I stood still, closed my eyes, and saw them both at once.

Storms of the Heart

When the rain finally came, it fell as if from the lip of an overturned bucket, then as though someone had tipped the earth and spilled the ocean. Wave after wave of water splashed down. My windshield wipers could not keep up with the overflow so I pulled off the road and sat listening to the hammer of drops on the roof of my car.

It's like that sometimes. The gray clouds hover, you expect the rain, but what you get is a deluge—more life than you can handle at one time. It further occurred to me as I watched the slide of water over the window that no matter what happens in nature, there's a mirror of human behavior somewhere in the event.

My own life weather has been tempestuous, with just enough calm moments in-between to suspect that there always will be more. The storms still catch me unaware, however, despite the warning signs—the rumblings of discontent like far off but audible thunder, the harsh words and the heavy silences that precede the tempest.

I was taught as a child to harness the lightning that threatened to blast relationships to smithereens in one brilliant, violent flash. I was severely reprimanded for pulling my brother's ear in an argument, was punished for biting and kicking and scratching to settle a score, was banished to my room for outbursts of verbal wrath.

Slowly but surely, I built a container for my anger, put up walls. I wore deliberate kindness as a suit of armor and learned that what you practice in earnest can become an integral part of you. There were always chinks in my armor, though, and I was continuously breaching the walls to see if the climate outside my own sense of self had changed.

As a teenager, I channeled my intensity and passion into safe outlets—pouring my turbulent thoughts into stories and diary entries and poems. I created happy endings where none existed, wrestled with fears, practiced acceptance, wrote and rewrote my real life into something I could believe in. I learned that I could be at odds with everyone else's real life and still be secure in mine.

As an adult, I've come to see the mirror in natural events as an aide to figuring things out. Just as surely as I knew the rain sluicing over my car and obliterating the highway would ease and finally cease, I know, too, that storms of the heart blow themselves out. The tears, like the rain, get blown away by the incessant winds that ride the sky, winds that change the world just by moving through.

I have learned that try as I might to outrun my life storms, they always catch up with me. Far better to stop by the side of the road and wait them out, letting the rain—and the tears—wash down, revive, cleanse, release, than to shake my fist in helpless anger at a natural turn of events.

Gradually, as I sat thinking these very thoughts, the rain ceased pounding, became a gentle patter. I eased my car back onto the highway, safely along a road I could now see.

Cathedrals and Grass Angels

We have had a spate of exquisite fall days, the kind that make the heart ache and the spirit soar simultaneously; the kind where the sun turns frost-wilted stalks of corn from gold to burnished copper and filters through the yellow poplar leaves until they glow like miniatures of the star that lights them. These are the days when maples hold up their leaves like crimson banners and flocks of geese stitch cryptic patterns in the fabric of the sky.

I went for a walk on such a day, when the light was soft and hazy and the smallest of breezes played in my hair. I stood transfixed and watched the leaves come spiraling down, their forms as light as fairy wings, their colors as vivid as though they'd been freshly painted. It took my breath away to be there in the spill of light with the leaves spinning about me in a joyous dance.

I found a thicket where blackberries had grown and ripened in the August heat. Beneath the rosy foliage I found two plump, forgotten berries and popped them in my mouth. Their unexpected taste of summer stayed on my tongue long after I had eaten them. I found, too, an old apple tree, its fruit small and twisted but nonetheless sweet. Fallen apples were thick in the grass at the base of the tree and there were deer prints in the soft earth.

Farther on, where the trees thinned and a wide expanse of sky stretched high and blue, I watched a hawk circle. Slowly it turned, its wings catching the updraft, its cruel beak and far-seeing eye visible from where I was. I lay on my back in the meadow and watched it circle lazily, wondering what it made of me, such a large and motionless intruder trespassing in its domain. I looked beyond the hawk, as deeply into the blue as my eyes could see, and thought about seasons and cycles and wind under wings, knowing that life's puzzles might not be mine to understand.

I moved on when the hawk did, making my way into the woods that bordered the meadow. Here the late afternoon sunshine filtered through the trees, casting splotches of leaf shadow that quivered when the wind stirred. The high, arching branches of the trees formed a cathedral where birds fluttered and sang and the muted light wavered like candle flames. I found the offering of a turkey feather, its edges ruffled, lying among the leaves, and made my own prayer for the beauty and peace I found there.

Before the sun spilled the last of its waning light, I made my way home through a field of cut stubble and abandoned corn cobs. In an uncut part of the meadow, I stopped and flung myself down on the feathery grass, breathing in the sweet scent, spreading my arms wide to embrace the sky. When I rose, I saw that I had made a grass angel; there in the meadow was my shape—round head, far-flung arms, long legs, impressed on the grass like a shadow. I left it there, undisturbed, to guard my place on this earth.

Signs to live by

At a local entrance to the Appalachian Trail, there are two arrows. MAINE says one, pointing north. GEORGIA says the other, pointing south. That's it. There are no further directions; where you go is determined by those two arrows and your own decision. There's no mention of where you'll end up if you go east or west, no warnings of bears, snakes, scorpions, inclement weather, or boulders in the path. You are not told to wear sunscreen, eat your vegetables, dress warmly, carry a flashlight. How you get there, what condition you'll be in when you arrive, how far along the path you go are all up to you. All the signs do is give you a directional option—north or south, Maine or Georgia?

I like those signs. They're clean and simple. They don't lose their essence in explanation or instruction—they just point. I've found, over the years, that my best experiences have come when I followed the simplest of ordinances, and those that came from an inner urging rather than an obedience to some other, outer voice.

In a store the other day I came upon a sundial. On its face, in raised letters, were these words—"Ignore the dull days, forget the showers. Keep count of only shining hours."

As far as advice goes, it doesn't go far. You aren't told how to, only that you ought to. It's an interesting concept, presented on a device that can *only* count shining hours and suggests that perhaps happiness is the art of finding joy in the little privileges of life. There's no danger of developing eyestrain if you look only at the bright side of things.

It also assumes that the principal business of life is to enjoy it. Were you to count just the golden days, remember only the good stuff, just think how happy you could be. But the sundial can't tell you how. It only points at a possible direction.

On the way to my daughter's home on Cape Cod, a huge billboard looms up over the highway. "Be a little kinder than necessary," it exhorts.

The first time I saw it, I thought, what a good thing to do—be kinder than necessary. The concept behind the words assumes that being kind is *necessary*; being kinder is a superlative act. It requires a commitment, a conscious effort to do more, to care more.

The sign doesn't give directions as to what kind acts to perform, nor does it explain how kind is more kind. It simply points out an alternative to something we already know, that it is essential to be compassionate.

The signs are there. We are constatnly directing ourselves. The trick is in knowing that all a sign does is point. Where we go from there and how we go are up to us.

WOW! Moments

You wake up one morning and the sun is shining through your window. Tiny little dust motes dance the dance of life in the light beam and the silence of the day's beginning is so loud you can hear it. Then a cardinal pipes his summer song from a nearby tree and the silence wavers a little. You hold your hand up to the light and see the blood pulsing through the network of delicate veins just under your skin. The silence and the song and the sight of your own lifeblood throbbing come together in a moment of intense awareness and you think, WOW! How can a day be ordinary after that?

We think of our days as commonplace but we're mistaken. Pick something, anything and really look it. What happens? The ordinary becomes extraordinary. I clasp my early morning mug of tea in my two hands and I think of two other hands that held raw clay and shaped it into the cup I now hold. I think of the discoveries that led to firing and glazing; I think beyond the mug to the tea inside it. What I drink so often (and so often heedlessly) was once an integral part of a living plant growing somewhere on a hillside a continent away.

I have a friend who is hiking the Appalachian Trail for the first time. She's never before pushed herself to walk more than a few miles, never spent more than several hours totally immersed in natural surroundings, never spent a night in the woods by herself. Now she is climbing, higher than she's ever been and the view of her surroundings takes her breath away. "It was there all the time," she says but now when she climbs a peak, turns a corner, looks out over vast tracts of unpopulated forest and woodland, "it's like, WOW! That's incredible!"

We need to do more than just stop and smell the roses. We need to immerse ourselves in the moment, every moment. "I can't," you say. "I can't oooh and aahhh my way through life. I'd never get anything done."

I say what we do now is only half done if we can't make the moment count, if it isn't considered a miracle. What is time but a string of moments, some noticed, some lost in the rush, as we hurry to "get things done"?

It doesn't take much time to listen to a bird sing, to appreciate the scent of flowers carried on the wind, to feel the warmth of a sunbeam. What better thing would we be doing with that fraction of a moment?

I saw a Baltimore Oriole the other morning. I haven't seen one in years but suddenly, there in the bush at the side of the road, like an unexpected gift, was a brilliant orange bird. It lifted into the air and glided across the meadow grasses as silently as a shadow, disappearing into the treetops as I watched. "Wow!" I breathed, thoroughly enchanted by that glimpse of silent, spectacular color in flight. The image has stayed with me for days. The memory of it makes me draw in my breath and smile, just as I did when I saw the bird in that fleeting instant.

All our days are made up of such moments. All of our moments are WOWs—we just need to be reminded now and then.

The Dreaming Bench

It is only twelve wooden slats on wrought iron legs but it supports a thousand dreams. The dreaming bench sits under a linden tree on the front lawn, facing west and the setting sun. Bob, the practical one in our household, rescued it from a junk pile, glued and bolted its broken leg and set it out saying, "I've always wanted one of these."

I had, too, but didn't know it until I sat on it. The slats are covered with worn and chipped brown paint. They curve down comfortably from the back to form a roomy seat big enough for two. The supporting legs are intricately formed with tendrils curving around a wrought iron rose nestled cunningly beneath each arm rest.

The morning sun rises just behind the dreaming bench, gilding the legs, turning the color of the wood to honey. I tiptoe through the dew-damp grass in bare feet and curl up on the seat in a pool of sunshine. The world around me, bathed in ethereal first-light, looks like a watercolor, all soft and blurry at the edges. The grass is painted in a wash of green and gold; bright yellow stars with deep brown centers will become black-eyed susans when the light grows stronger. Now they are nebulous shapes in the mist, bits of heaven-dropped color at the edge of the garden.

All around me there is silence, broken only by an occasional bird twitter. Here I fashion my morning dreams, letting my thoughts go slowly into the day before me. It is a place to gather the silence and the beauty that will underlie the day's activity. I dream my day the way I would like it to be, full of peace and unhurried harmony. When the pace becomes hectic, I can pull out the flower of my morning dreams and place petals of tranquillity in the frantic moments.

Under the hot noon sun, the dreaming bench is a haven of shady comfort. If I'm lucky, it is a place to eat an unhurried lunch and sip iced coffee through a straw. The rush of noonday traffic lies close by on the road and I can wave leisurely at the drivers as they hurry past. When my own pace is frenetic, a few snatched moments on the dreaming bench can be as refreshing as an oasis.

Late in the afternoon, home from work, with the slanting sun reaching long fingers across the lawn, I take a mug of tea to the dreaming bench and watch the butterflies sip nectar from the garden flowers. A turn of my head gives me a glimpse of the blackberry patch where I'll soon be, picking warm, sun-ripened berries for a supper pie. Near it is the vegetable garden where onions and lettuce wait to be turned into salad. I relax in the warm, subdued light, and let my swirling thoughts wander and evaporate with the steam rising in fanciful eddies from my tea cup.

Evening comes. Supper has been eaten, the dishes done. There is nothing left of the day but these last few moments of fading light turned crimson by the setting sun. The dreaming bench waits beneath the linden, waits for me to come and dream my dusk colored dreams. In the waning light I ponder the meaning of circles and spirals, of the day gone by and the day to come. When twilight falls and the distant light of far-flung stars reaches my eyes, I can only wonder at my presence here.

Rain Reflections

Sunny days yank me up out of bed early in the morning. *Get up, come out,* they beckon and I can't resist. I like to watch the sun splash its way across the meadow, kiss the treetops and spill its light across the yard. I want to be out in all that brightness, bustling and busy. Sunny mornings are energizing.

Rainy days, on the other hand, are an invitation to snuggle under the covers just a few minutes longer. They say, *slow down, slow down*, coaxing me to linger awhile over a second cup of coffee or stand idly at the window watching the rain fall in gray sheets from a grayer sky. Rain hypnotizes, drum-drumming on the roof, drip-dripping from the eaves.

I notice that my tolerance for rain changes with the seasons. Autumn rains seem dreary and monotonous. They fall from leaden skies, draining color from the landscape, erasing the tapestry of color woven by dying leaves. The soul knows they herald the beginning of the bleak season.

The winter rains that follow can be both dismal and exhilarating. Bitter, freezing rain is most discouraging, trapping me indoors, threatening my safety if I venture out. In contrast, the blustery rain of a January thaw is a marvelous thing to be out in. It is elemental, rage in the form of wind and water, and eventually a relief to come in out of. I walk in it but not far. I stand in it but not for long. I watch it erase the snow, uncovering the bones of the land, exposing bits of sodden corn stubble and wind-flung tree branches. Rain in winter is always a challenge, flung down at the feet with vehemence.

When finally winter eases and the cold must, of necessity, give way to the insistence of the sun, the rain again changes its tenor and falls as April showers. The tempestuous, wind-driven storms of winter blow themselves out, leaving in their wake the gentle, steady rains of late spring. The skies are still drab and discouraging but the earth responds to these greening rains, putting forth shoot and blade, bud and leaf, until the whole earth sings with color despite the gloom. Rainy days often outnumber sunny ones as spring seeks to establish itself. No wonder the day that dawns warm and bright pulls me up and out, to shake winter from my bones and open my soul like a flower to the sun.

Summer rains are changelings. I have been caught in both brief, violent storms, rife with thunder and lightning or the slow, warm, soaking rains that gardeners and fishermen love. Often during summer, there comes a day when the sun refuses to shine and the sky is the color of an old metal washtub. Rain falls slowly, soaking into the green earth, washing clean the leaves and grass blades and flower petals and the stones at the side of the road. These are the nourishing waters that slake the thirst of growing things.

It is raining now, as I write—a long, slow, steady fall of water. It is still early May and the temperature hovers only in the fifties. I will fetch an umbrella and my boots and go for a walk, letting the gentle shower speak to me of sunny days to come.

Rainbows and Imagination

A co-worker retired last week. At his party he received the usual watch and well wishes. He looked a little wistful, though, as he thanked us all and reiterated how much he was going to enjoy his freedom. To ensure that he did, I bought him a huge jar of bubbles and presented him with it on the last day of school. He looked at me quizzically.

"Bubbles?" he said.

"Rainbows," I replied.

I am a firm believer in the power of bubbles. They start out as ordinary soap and water. But catch the soapy film on a wand, blow gently, and look! You've created a rainbow. A bubble is as delicate and ephemeral as an angel's wing, a perfectly round, incandescent sphere that reflects its surroundings as it floats effortlessly on the air. Even on a cloudy day, bubbles glow with rainbow colors. Seen in the sunlight, they dazzle the eye.

Everyone needs rainbows. Children need them as guides to the fulfillment of their potential. Adults need them as a reminder of the beauty of life. Too often we get bogged down by the weight of our own thoughts, confusing seriousness with importance. What's important can be found in a bubble. It is at once delicate and strong, borne aloft by the slightest breeze yet capable of landing so gently on a sharp object that it does not break. It is a thing of unutterable beauty yet it has its beginnings in common soap and water.

Watch a child blowing bubbles for the first time. It is a sloppy affair. Soap drips down the fingers, leaves trails on a dirty arm. The lips purse. The first sharp breath blows the film to smithereens. But gradually the child learns to dip the wand carefully, to control the force of air. A bubble begins to grow at the end of the wand, an instant rainbow that floats off ever so gently, up and up and up. When the beautiful thing bursts, the child is crestfallen. But there in his hand is the jar, still full of rainbows waiting to be launched.

No adult should be without a jar of bubbles. I resort to mine often, as a pick-me-up for low spirits, as a reminder that from lowly beginnings come great things, as a balm for sadness and a celebration of beauty. I sit on the rock wall or stand at the garden's edge and send my rainbows far and wide.

Life's Lessons

Before someone told me the awful truth, I was content to view the world as consisting of two bowls, like the china ones in my mother's cupboard. One bowl was filled with earth, upon which some thoughtful deity had planted grass and flowers and trees. Atop this, upside down, it's inner sides painted the loveliest blue, rested the other bowl. It was a comfort to know that when I lay down on my bed and pulled the covers up to my chin, I was safely ensconced between the two bowls and that I would never, ever fall out.

Eventually, some well-meaning adult told me that I actually was standing on the outside of a huge ball in mid-space and worse still, the ball was spinning. Gravity, it was explained to my horrified mind, was the force that pulled all objects earthward and kept them from flying off into space. It was hardly reassuring, now that my bowls were smashed to smithereens. It was a long time before I felt safe again.

I learned other frightening truths as I grew up. One was that nothing and no one, no matter how well loved, stays with you forever. My first brush with death occurred when I was five. It was Sunday morning and we had just returned from church. Still wearing my dress-up clothes, I went looking for my new kitten only to discover it crying piteously on the doorstep, half in and half out of the mouth of a snake.

Appalled, I screamed for my mother. She dealt the snake a blow with the garden shovel but it was too late. My poor little kitten died moments later. I was inconsolable. Why, I argued to my mother's explanation, could not God have gotten his own kitten? She had no answer and I began my long journey away from the simple faith of my childhood and into the world of the unfathomable.

I was fairly selfish and self-contained, as children are wont to be, and I was sure that what happened in my world happened in everyone's. When I stood at the top of the stairs in my pajamas and implored my mother to come up and get the monsters out from under my bed, she told me there weren't any monsters. She didn't even come to look. She would have seen them there, as plain as anything, their long tails curving into the darkness under the bed, their evil eyes casting reflections in the window.

I turned my bedroom light on and that brought her up the stairs. I told her that monsters didn't show up in the light but she didn't believe me. She turned the light out with strict instructions to keep it out and left me alone with my demons.

I was right about that, though. All my monsters still disappear with the dawn. In the dark reaches of the night, when my mind turns somersaults, when undone tasks loom large or regrets come to haunt me, I remember that in the light of day these things will take on their ordinary, unthreatening proportions and the world will again resemble two bowls of safety. I breathe easier and sleep tighter, knowing that even the tiniest light cannot be defeated by darkness.

Friends

includes:

•Woods Walks and Apple Pies
•Junie's Garden
•Saying Goodbye
•Growing Up, Not Old
•The Best of Bernie

Woods Walks and Apple Pies

Driving into the hills of northern Vermont I think, this is what an ant feels, exploring the depths of a rose. I-91 is the ant path I take through the mountain petals, each one rising up from the center to curve back in on itself or fold back to reveal ever more hills. In every direction I see mountains.

I go to Vermont to visit my friend, Lora, who lives in the Northeast Kingdom, four hours due north from my home in western Massachusetts. She is a Vermont native, having lived all of her 91 years in these mountains and meadows. Her house overlooks a lake nestled among wooded hills that shimmer in reflection on the water's still surface. We take a walk to the lake and skirt its edges, passing the camps where summer people come to breathe the clear mountain air and spend their vacations fishing and swimming.

At this time of year, the bungalows are shuttered and vacant, with a forlorn and desolate look about them. Soon the leaves, having lost their chlorophyll camouflage, will drift down like scattered jewels to lie in drifts on the unkempt lawns. Winter will come to pile snowdrifts high against the foundations and atop the roofs, and the lake will be trapped in two feet of ice. Finally, spring will turn these woods roads to mud before summer comes again to liberate them.

Back at the house, we scavenge Lora's garden for enough ears of sweet corn and tomatoes for our supper. Her garden is large and her two freezers are packed with homegrown vegetables. We eat like there's no tomorrow and on a farmer's schedule—breakfast just past dawn, dinner at eleven-thirty, and supper at five.

On one day we drive to a neighboring farm to pick apples, bright red globes that cling to branches way over our heads. Lora drives her pick-up truck under the tree and I clamber up onto the toolbox at the far end of the truck bed. I reach the apple picker into the branches and shake. Apples rain down, peppering our heads and the truck before rolling into the deep meadow grass or splatting in the dirt beneath the tree. A few remain in the picker's basket. We put these unbruised fruit aside for eating out of hand and fill two bushel baskets for pies and sauce.

The promised rain holds off and we go for a ride, stopping first at a wagon road that leads to a small pond in the woods. Vermont is full of such places that harbor trout and perch in their cold waters. We scuff at a few fallen leaves and catch glimpses of blue between the tree trunks as we near the pond.

I feel as I have often felt before, that here in the company of trees and wind and sunshine I have all I need of beauty and pageantry. No man-made scenery can surpass what it can only mimic. Deep in the woods, my eyes, my ears, my soul are filled so completely that all else is merely extra.

On the road again we stop to inspect a camp for sale that has caught Lora's eye. Further on, we come upon a pond monster, a rival of the serpent purportedly lurking in the depths of Lake Memphremagog. This one is made of pumpkins and squash, with ears of corn jutting from its head. The creation of Representative Robert Kinsey of Craftsbury, the monster, dubbed Ally Looya, floats serenely on Bob's pond. An intrepid photographer has snapped its picture for a local newspaper to alert the public of the monster's presence.

Another day, Lora and I and Barry, Lora's nephew go looking for moose in the early morning hours. Through the rising mist the sun looks like a dot of melting butter in the sky and the trees are shrouded and mysterious. When we drive out of the fog into the brilliant sunshine, the blue-green spruce, the birch with its spotted and peeling bark, the yellow poplars, and the gnarled maples look again like ordinary, every-day trees. We gave no prior notice of our visit to the moose bog and they have obviously made other plans. We did not see a single one. We console ourselves with a large breakfast in a small restaurant before driving home.

Evening falls earlier now that the earth has turned another season. Together Lora and I watch the sun sink slowly behind the mountains, pulling curtains of darkness across the stage where we've played out our day. At night, I dream of moose and woods walks and apple pancakes browning slowly on a hot griddle just before dawn.

Junie's Garden

'So grieve for me awhile, if grieve you must,
then let your grief be comforted by trust.
It's only for awhile that we must part,
so bless the memories that lie within your heart.
I won't be far away, for life goes on,
but if you need me, call and I will come...'
Anon.

The last time I saw my friend Junie Cartinelli, he was doing what he did best—he was helping. I'd called to ask if he'd talk with me about bee keeping for an article I was writing.

"Sure," he said. "Come on over," and so I did.

Piled high on his kitchen table were magazines about bees, order blanks for bee keeping supplies, pictures and articles and books on the subject.

"I don't know if I can help you much," he said earnestly, waving a hand at the piles "but all this might."

That was Junie all over. I never once heard him boast about what he knew, and yet he seemed to know something about almost everything.

I'd known Junie almost all my life. He was the much older brother of one of my best friends, married and having a family of his own by the time I met him. He kept bees and my father would take me to his house sometimes to get honey. I loved to go. The house, the gardens, the fenced pastures were neat as a pin. Junie would put the box of golden honeycomb in my hands as though he was certain I was old enough and careful enough to be handed such a treasure. I had a little girl's admiration for this strong, rugged, handsome man with the gentle heart, and over the years he became my friend.

Junie was a carpenter and builder by trade though he would rather have been a farmer. No one who saw the well manicured acres of his small farmholding or watched the care he lavished on orchard, garden or animals could doubt his love for the land.

Nor did those of us who knew him ever doubt his love for his fellow man. He raised vegetables, sold eggs, kept fruit trees, and grew flowers—to the benefit of every one of his friends and relatives. No one left Junie's house without some gift—an armful of corn, a bouquet of flowers, an apple or two for the road.

He dispensed advice and friendship as readily as he did garden crops. He believed in hard work, in doing one's best, and in taking care of ones' own. With Barbara, his wife of forty-two years, he raised three children, and when his only grandson was born, he took him under his wing and into his heart with the rest of us (not to mention into the woods, the barn, and the garden). In fact, Junie took such pride in his home and family that his love and enthusiasm spilled over into every other aspect of his life. We who knew him were the lucky beneficiaries of his largess.

A couple of summers ago, he appeared at my doorstep with a spackle bucket full of cucumbers.

"I didn't know if you had enough of these to make pickles," he said, setting the bucket down on the porch. "I thought maybe if I shared the cucumbers, you'd share the pickles."

I peeled and sliced and packed several dozen cucumbers into canning jars. When I was finished, I had two dozen quarts of bread and butter pickles. Junie would only take six. "It's all I can use," he said, and as usual, someone else got the better end of one of Junie's deals.

This year, with the last of winter washing away in the spring rains and his garden just waiting for the spade, Junie became suddenly ill. He died so quickly that most of us were taken by surprise. Several weeks later, I went by to see Barbara, to talk with her about Junie and to reminisce. We sat at the kitchen table and it seemed that at any moment he would walk in. But he didn't, and as I was leaving, Barbara stopped me and pointed to the side yard.

There, lit by the last rays of the setting sun, was Junie's garden! Row after neat row, growing green and sturdy, were the vegetables Junie's children and grandson had planted for him.

It would have pleased him greatly to know that his faith in nature was justified, that death is only life beginning in another form, and that the love he showered on family and friends and nature itself has been plowed under and is growing up again.

Saying Good-Bye

There's a fairy ring in my yard—a circle about 11 feet in diameter composed of tiny white flowers, with a smaller circle in the exact middle where no flowers grow. When all of my human machinations fail, it is where I go to make wishes.

Tonight I went there to make a wish for a friend. She is dying. After battling cancer for several years, her body is finally succumbing. She is awake and lucid for only a few minutes at a time; her whole self is making preparation for the swift change it knows is inevitable.

"Just a couple of weeks," the doctor told her at her last visit. She packed a few things, left her house in Florida and traveled to Vermont to say her good-byes to the family and friends she'd left behind when she moved. She will be buried there among the green mountains when she dies, close to her children, finally a part of the land she's loved all her life.

The wish I made was not that she go on living, though I asked for that first. I was looking at her death through my own fears, feeling panicky at the thought of her having to leave behind the dear and familiar. I remembered too clearly my father's death, how he resisted dying, how he wept at the thought of leaving us, how he struggled to stay and could not. I recalled, too, my mother's farewells to my children; the tears in her eyes, and theirs, (and mine) as they said good-bye for the last time. There was such sadness, such finality, that it haunts me still.

Once through and on the other side of that panic, I wished instead that this friend feel, as she let go of consciousness, completely surrounded by all the love she's been doling out so generously for so many years. For this woman—daughter, friend, mother, wife—was determinedly cheerful in the face of adversity, encouraging when things were at their bleakest, always ready to laugh, and willing to forgive. She loved her family and her friends and she loved life.

I sat in the fairy ring and wished her god-speed, wished that she could take with her the memory of the setting sun that illuminates the whole world with gentle light; that the memory of rain would wash over her; that the color and scent of spring flowers would fill her senses; that the sweet sound of bird song, the delightful laughter of children, the lonesome whistle of a distant train echo in her ears long after she needs them to hear.

I wish she could have stayed longer. The day of departure has come too soon. It always does. No matter how prepared we are for the end of something, the pain of the final parting is a surprise. The great, aching emptiness swallows us whole. But after awhile, with the minutes and hours, the days and weeks and years following loss, we build a ladder of time and begin the climb up from despair to hope.

I won't know if my wish for her comes true. I won't know until it is my own time to die what lies ahead for her now. I make another wish in the fairy ring—that her death mean as much to her as her life.

Growing Up, Not Old

"I want to be just like you when I grow up," I told my friend Lora the other day. We were walking side by side on treadmills at the Coachworks Wellness Barn in Albany, Vermont. Just five months shy of 91, Lora has joined the fitness center so she has something to keep her occupied during the long winter months. And true to form, that something might just as well be productive.

Upbeat music was playing over the speaker system, a cold November wind was scattering the leaves across the road outside the window and the two of us were taking a stationary hike, something Lora does at least three mornings a week now. She also lifts weights, sits on, stands in, and hooks herself up to the latest workout contraptions and swims a lap or two in the pool before heading back home for lunch and a nap. I was hard pressed to keep up.

She gave me a tour of the spotlessly clean place, coaxing me to try the various machines. With the aid of trained club supervisor Sue Grimes, I stretched (briefly) my arm muscles, lifted weights with my legs, and tried a free weight exercise.

We inspected the heated lap pool where Lora told me she'd learned to swim with her face under water.

"I was so pleased with myself for daring to do that, I thought 'Nothing's going to stop me now,' and I traded in my pickup truck for a 4-wheel drive Jimmy so I can drive over here in the winter. The snow won't stop me now!"

I thought wistfully that perhaps by the time I turned 91, I'd have some measure of Lora's perspicacity. I also wished I'd dug my bathing suit out of winter storage before heading north.

After the tour, we wandered into the delightfully appointed lobby. Books on fitness—of both mind and body—lined the bookshelves, a gas fire burned cheerily in the fireplace, a tray of fresh cut fruit and a carafe of iced tea waited on the counter. We sat in easy chairs, sipped our tea, and remarked to each other how lucky we were to be healthy and happy and no older than we thought we ought to be.

Most visits to Lora end up with me discovering some new way to approach life, usually involving a head on collision with some cherished belief I've been carting around for safety. Under her tutelage, I'm practicing to be happy even in the midst of anguish, to never think I'm lost but to assure myself I'm just taking the long way home, and to not cling so tightly to the past that I keep myself from stepping bravely into the future.

Going to the Wellness Barn is just another of Lora's ways to squeeze the most out of the moment. She even contributed a piece to the Coachworks newsletter, a smattering of comments on old age by an unknown author that starts, "It is twice as far to the corner now, and I notice they've added a hill."

Hills won't ever stop Lora, nor distance either. We climb back into her new red Jimmy and head for home, her body and my mind stretched and beginning to be a lot more supple.

The Best of Bernie

My friend, Bernie Atherton, lived for 94 years in the rugged hills of northern Vermont. He was already 44 years old when I was born and 80 before our paths crossed. He and his wife, Lora, were living on a farm in Greensboro when I first met them. Bernie reminded me of St. Nick—the round, jolly version minus the beard. There's a picture of him on the living room wall, a sturdy, stalwart farmer holding a small spray of daisies in his hand, an impish grin on his ruddy face. That was Bernie—concurrently strong and soft.

By the time I was getting acquainted with them, Bernie and Lora had sold off all the cows and the farm equipment. They were retired, although that's a misnomer when applied to either of them. Bernie didn't like to "just sit", so he spent hours designing and building bird houses and more hours watching all the birds that frequented them. He made picture frames for Lora's paintings. An avid collector, he scouted tag and yard sales, searching out interesting bottles and old glass insulators. The last few years he collected buttons, sorting them and affixing them to cardboard mats. He mounted 3,500 in all.

Besides helping Lora raise and preserve all the fruits and vegetables they could eat, Bernie grew gladiolas—fragile blossoms in a rainbow of delicate colors. He loved them for their beauty and gave them away to anyone who admired them, especially appreciative women. A friend of his who owned the little country store in town chided Bernie's nephew once.

"You should take a lesson from your Uncle Bernie and give flowers to all the girls," she told him.

"Oh, I don't know," was the nephew's reluctant reply. "Old men can get away with things young men can't."

Bernie could get away with a lot. He was forever teasing. When I went to visit them, Lora and I would spread our paints out on the kitchen table and spend the afternoon creating masterpieces. Bernie would cast a critical eye on our attempts.

"What do you think, Bernie?" I would ask, knowing I shouldn't have.

"Well, sister," he'd say with a twinkle. "I have a barn out back that needs painting," and he'd go off giggling.

"He can't let a thing pass without some comment," Lora once remarked. "If I was the one backing the hay trailer into the barn, he'd tease me about taking three tries to his one or if I was raking the hay field he'd shake his head and wonder aloud how I always managed to skip the corners. He knew just how to push my buttons," but she smiled when she said it.

Bernie had his own vernacular, a mixture of Vermont-ese and Bernie-isms. He'd intersperse his comments with "don't cha know," and "by crimus." If he was describing something, he'd always add, "and like-a that," to the end of his sentences. Bernie never used swear words to make a point. Judas Priest! He had so many substitutes he didn't need to.

Every spring that I knew them, he and Lora engaged in what became known as the Great Potato War. Bernie insisted (by crimus) that potatoes had to be planted the last week in May (dont cha know). Lora felt that any way she could get a jump start on the potato crop was acceptable. So, they each planted their own potatoes and all summer long they argued about who had the best spuds.

Bernie didn't have much time to gloat over his win this year. Not long after he and Lora put their garden to bed in the fall, pulling up the dead stalks and frostbitten vines, Bernie took ill and was rushed to the hospital. Lora called me with the news of his death a day later. I rode the long, upward miles from Massachusetts to Vermont, wondering at the cycles that move us in and out of other people's lives like seasonal winds. When I got to their house, the family was gathering. We talked of Bernie late into the night.

Said Bernie's nephew in summation, "He was probably the most successful man I knew. He had all he needed for the lifestyle he wanted—work he liked, great friends, and the respect of others. He married the girl he loved and she stayed with him for sixty-seven years, loving and caring for him. He had a long and happy life."

What more could any of us ask?

Family

includes:

•Beacuse of Them
•Picky Fwadoos
•Can-Do Lessons From My Son
•Love Letter to My Daughter Who is Far Away at Christmas
•A Letter to My Daughter as She Graduates
•The Gifts of Children
•Thr Great Sled Race
•Freedom Ride
•Lick, Spit and Polish
•Missing Dad
•Remembering Mama

Because Of Them

This past Sunday I celebrated my twenty-eighth year as a mom. Brendan, my first-born, arrived on Father's Day—an appropriate gift. He was a tiny, pink, frowning replica of his father. His head was covered with a thatch of golden curls and his eyes had the intensity of a summer sky. He would gaze at me so long and questioningly as he nursed that I would have to look away or cry. I loved him wildly. I still do.

He was an inquisitive child, and a happy one, until the first day of school. I opened the front door at mid-morning to find him crying on the doorstep. He had walked the scant mile home during recess. His teacher called in a panic when she couldn't find him. As I walked him back to school he explained to me solemnly, "Six hours is far too long for a child my age to be away from his mother."

No kidding.

When he was seven, I put him on a bus to Boston with a chum from school. They were going to visit the friend's grandfather and Bren was beside himself with excitement. I was beside myself with worry. In his suitcase I packed Sammy Sock, his favorite toy, thinking that it would help him be less homesick if he had something familiar with him. When he came home two days later he confided that he had been having a wonderful time until he opened his suitcase to get his pajamas. Seeing Sammy made him so homesick he couldn't sleep! It was an important lesson—never presume your child feels the way you do.

As he grew, he seemed to stretch. Always thin, he soon towered over me like a young sapling. I remember a day I was scolding him over some trivial thing. He looked down fondly at my upturned frown and said, "What did you say, Shorty?"

Incensed, I dragged a chair over and stood on it, forcing him to look up at me. The absurdity of it struck us both at the same time and we collapsed in a giggling heap.

My second son was born a year and a day after his brother. For the first two days of his life he was know as Baby because his dad and I couldn't agree on a name. Finally in desperation I began reading aloud from *What To Name Your Baby.* When I got to Kenneth, I looked down at Baby. He looked back and gave a ghost of a smile. I didn't care what we called him as long as I got to keep him. I loved him wildly. I still do.

Ken was a rugged little boy, with a mop of yellow curls and a mischievous grin. He liked to move fast and climb high. When he was two I went into the kitchen one morning to find him sitting Buddha-like on top of the refrigerator. When he was five a neighbor called in a dither because she'd seen him climbing among the branches of a huge pine tree in her yard. I strolled over to fetch him.

"Hi, Mom," he called from his hiding place twenty feet up.

When he was old enough to get his driver's license his standard excuse for a speeding ticket was, "But my car just goes too fast!"

Jennifer was born two months after Ken's first birthday. Her hair was as soft as swan's down, her eyes the color of violets. Her brothers adored her. I loved her wildly. I still do.

She was enchanting—a butterfly child who sang to herself every morning while waiting to be rescued from her crib. One

Christmas my mother came to spend the holiday with us. She brought a miniature Christmas tree trimmed with blinking lights for the kids' room. Before going to bed she plugged the cord into the socket. She woke to find Jennifer sitting on the foot of her bed, watching the lights blink on and off.

"Isn't it pretty? Isn't it nice?" piped Jen. We still say that when entranced by some unexpected beauty.

"Watch me dance!" she cried at age seven, pirouetting around the dance studio.

"Watch me twirl!" she sang, throwing her baton high when she was twelve.

"Watch me do cartwheels!" she yelled from centerfield during half-time.

"Hear me sing!" she exclaimed, standing in front of the microphone at her high school graduation. She was never so happy as when she was on center stage. Where did this extrovert come from, I wondered?

Cassie, the caboose baby, was as sturdy as Jen was elfin. She had the same golden hair and wide blue eyes shared by her siblings and a deep chuckle reminiscent of Greta Garbo's throaty tones. The day she was born my father stood over her, examining her fingers and toes.

"Look," he said, "at her perfect little finger nails. Look at her perfect little feet. Every time, it's a miracle."

I loved her wildly. I still do.

"Don't see me, Mommy," was her favorite admonition as she was growing up. She was an intensely private and self-reliant child. "Born old," people said of her, and it was true. She seemed to possess some ancient wisdom that gave her an enviable self-assurance. When she was thirteen I put her on an airplane to

California where she was to meet my sister. From there they would fly "down under," spending the summer backpacking through Australia and New Zealand. I got dizzy every time I imagined the soles of her feet meeting mine from the other side of the earth. *Dear Mom,* one bloodstained letter began. *I can write now that my cuts are almost healed...* Of the two of us, she was (and is) the one with admirable aplomb.

As a girl, I feted my own mother with flowers and perfume on Mother's Day. My own children, flesh of my flesh, nourished by my heart's blood, are my reason for celebrating Mother's Day now. Because of them my life is enriched beyond measure. I loved them wildly. I still do.

Picky Fwadoos

Spring, twenty years ago, and a small hand is tugging at my sleeve.

"Picky fwadoos, Mommy," begs my firstborn and out we go to pick the flowers that have blossomed in the warm sweet air.

I remember this as I rake last year's leaf mulch from the garden bed. Crocus—yellow, purple, and creamy white—hold their cupped flowers open to the sun. *Picky fwadoos*. I pluck one of each color for the cobalt vase.

As a child myself, I named every flower I saw, never guessing that someone else had done so before me. In my eyes, yellow trumpeted daffodils and their cousins, the narcissus, hearalded the coming of the king and queen of the garden fairies. Hyacinths scented the air and tulips bowed before the breezes. What heart can resist such magic? When the meadow beyond the house filled with field flowers, I would spend hours there day-dreaming in the warm sunshine. It's still one of my favorite pastimes, dreaming among the daisies.

I remember picking a sampling of all the flowers I could find to take home to my mother. Spreading them out before her, I would name every one. My mother, ever the school teacher, would fetch the big, yellow covered flower book from the shelf and pronounce the Latin names of each of my specimens. I don't remember a single one but ask me what a queen's cup is or a golden trumpet or an orange popper and I'll show you.

My toddler had the same independent spirit. When I told him the name of the forsythia bush that blossomed in golden profusion at the edge of the road he asked, "If Cynthia has a flower with her name, is there a for-Brendan flower?"

I never see one of those bushes in bloom without thinking of my son. It's a lovely reminder. Now, when the first flowers are pushing their way up out of the cold earth, when the pale green spears of the lily and the round, raggedy edged violet leaves promise sweet blossoms, I go out to "picky fwadoos" in memory of the little boy who reopened my eyes to the wonder of springtime.

He's a grown man now, and lives far away but I know somewhere in that tall, lanky frame there lurks the small son who will pick spring flowers and think of me, too.

Can-do Lessons From My Son

He was only two when I discovered him sitting Buddha-like on top of the kitchen refrigerator. To this day I have no idea how he got up there all by himself. I stood on a chair and lifted him down. "No, Mommy," he pleaded. "Up!" and he squirmed to be set free.

When he was five, he learned to ride a bike with training wheels. He sped up and down the driveway, careening around the car on one training wheel while he made screeching tire noises through his lips. His dad finally took the extra wheels off.

"Wait a minute, Ken, while I put the wrench away and then I'll help you ride your bike."

The words weren't out of his mouth before Ken flew by, pedaling down the drive and out onto the road. His father stood shaking his head. "Look, I'm going fast!" Ken yelled.

Up. Fast. They were two of Ken's favorite words. It was a lesson his dad and I were both reluctantly learning—nothing held Ken back or slowed him down.

Just weeks before, we'd been swimming at the town pond where a lifeguard sat high on her wooden chair, surveying the water as dozens of children splashed and yelled. Only those who could swim under water were allowed to go beyond the closest rope to the larger swimming area beyond. The lifeguard repeatedly blew her whistle at Ken as he lifted the rope and ducked under it.

"Little boy," she said, finally exasperated. "You can't go past that rope unless you swim under it."

Eyeing first the girl and then the rope, Ken suddenly plunged head first into the water and emerged several feet beyond the rope. The lifeguard, who'd jumped to her feet, applauded him from atop her chair. Ken grinned and dog-paddled off.

I watched my small son and wondered where that fierce independence and courage came from. Certainly not from me, I thought, watching with my heart in my mouth as he splashed happily in the deep water. He was teaching me to let go far sooner than I wanted. I could feel resistance struggling with admiration for that indomitable spirit.

When he was a teenager, Ken and his close friend Mark took a bicycle trip from northern Vermont to Bridgeport, Connecticut where they boarded a ferry for Long Island and a week-long visit with Ken's grandmother. With heads together over a map spread out on the dining room table, the two boys planned the trip, calculating mileage, overnight stays, and stops for supplies.

The day they left, I stood on our hilltop and watched them pedal off down the driveway. When Ken turned to wave, I waved back. I would not weep, I thought fiercely. I would not worry about accidents or strangers on the road or vicious dogs. I would see Ken the way he saw himself—capable, resourceful, inventive, efficient.

More than once, that can-do attitude of his spurred me to take more chances myself, to lighten up, to see the possibilities of a predicament rather than the problems. From Ken, as from all my children, I've learned some of life's most important lessons.

Now grown to manhood and off on his own, Ken comes home to visit when he can. I stand and wave when he leaves, seeing again the small boy whose favorite words let him climb high and go far.

Love Letter to My Daughter Who Is Far Away at Christmas

The little people—that's what your grandfather used to call you. Nothing pleased him more than to know the little people were coming to see him. Now the four of you are grown and on your own, scattered across the country from one coast to the other, and nothing pleases me more than knowing you are coming home, if only for a short visit. You are the first to be married and your own home is far from here. You will not be making the journey to be with us for the holidays.

It is almost Christmas Eve. The tree stands in its customary place, waiting for bits of colored glass and tinsel to work their magic. Each ornament I lift from its nest of tissue paper evokes some memory of Christmas past. Here's the small blue angel that your grandmother, my mother, gave you when you were just a toddler. With your gold-spun hair and big blue eyes, you looked like an angel yourself.

"Let *me* do it," you insisted, hanging the trinket on the highest branch you could reach.

Trimming the tree is a task for children. How your eyes sparkled, reflecting the lights your big brothers so carefully strung among the branches. I lift a small elf out of the box, a homemade dough ornament given to you by your fifth grade teacher and think, *you should be hanging these on the tree, Jennifer.*

Early Christmas morning I will get up before everyone else and turn the tree lights on, remembering the morning so long ago you sat in your bunny- feet pajamas and gazed at the twinkling lights, saying over and over again, *"Isn't it pretty? Isn't it nice?"* I will whisper the words softly to myself and think of you.

For more than twenty years you made Christmas cookies with me, decorating them with gobs of colored frosting and sugar sprinkles. Each year you became more adept at rolling and cutting and decorating. On each gingerbread man I make alone this year, I will put a big frosting smile in memory of the little girl who helped at my side.

I hum along with the carols playing on the radio and remember how we sang aloud every Christmas song we knew as we baked or wrapped gifts or marched from store to store in search of the perfect present. How excited you were on Christmas morning when you and your brothers and sister saw that Santa had come in the night. To this day, there's still a gift "from Santa" for everyone under the tree. This year your perfect present is on its way to you. I can picture the look on your face when you open it and will hear the echo of your voice—"*Oh, thank you, Mommy!*"—across the miles.

Isn't it odd that a heart can ache and be joyful at the same time? We will all sit closer together at the table so your place won't look so glaringly empty. We will take turns talking to you on the phone, wishing you a happy Christmas. And I will wish this for you, my daughter—that all the joy you've brought me through the years be returned to you a thousandfold.

A Letter To My Daughter As She Graduates

Dear Cassie,

Commencement. The word means beginning, genesis, advent, dawn. Out there, beyond the familiar boundaries of home and school, lie the somedays of your dreams. You walk toward, and in doing so, you walk away from. That's the way of things. Here are some thoughts to keep in your heart as you travel your own path.

In all your thoughts, and in all your acts, in every hope, and in every fear, when you soar to the skies and when you fall to the ground, you are holding the other person's hand.

A.A. Milne

Always remember that you are loved. You were such a sweet little child, with your big blue eyes and flaxen hair and that deep, throaty chuckle. You held my hand when we went for walks and your talk was full of whats and whys and whens.

"What comes down in Spring?" you asked, watching winter's snowflakes spiral to the ground, and, "When will I be big?" as you watched—with longing—your older siblings climb into the big yellow school bus. I would have held you small and needy, but my heart knew better. Now that you have become a young woman, the time has come for me to let go. It is time for you to make your own way.

You yourself must set flame to the torches which you have brought.

Anonymous

Remember that you have a purpose. You may not be sure what that purpose is even though you knew you wanted to be a teacher when you were six, and a teacher still when you were seventeen. Crowd into your days every possible experience, every adventure. Live your life to the uttermost. Grow as much as it is in your power to grow.

You wake up in the morning, and lo! your purse is magically filled with twenty-four hours of the unmanufactured tissue of the universe. It is yours...

Arnold Bennett

Pay attention to the present. All the days leading up to this day make up your past. The future—all that lies ahead of you— begins now. The fleeting moment inbetween is, in reality, eternity. When you were small, you knew instinctively how to focus on whatever you were doing—making mud pies, dressing your dolls, coloring a picture. You learned to lose yourself in a book or a daydream. You poured every ounce of your being into the moment at hand. Distractions will come your way. Remember to spend your hours wisely each day and tomorrow will be taken care of.

Home is the place where, when you have to go there,
They have to take you in.
I should have called it
Something you somehow haven't to deserve.

Robert Frost

Home is also where the heart is. You've had several homes—the little house in Connecticut that you left when you were two ("Dood-bye room," you said as we closed each door behind us); the rented house high on the hill in Walden, Vermont, where you used to catch "snowdrops" on your tongue when the wild wind brought one snowstorm after another; the log cabin in Danville that you helped to build; the houses on Barnum and Silver Street, and finally, the dormitory at college where you've spent the last four years. Now that you are setting off to far places, remember this—as long as we are in each other's hearts, you are never far from home.

Love,
Mom

The Gifts of Children

Mom!" came the excited voice over the telephone. "Did you see the line up of planets?"

I set the phone down and ran out to the porch. The sky was blanketed in clouds. "Not tonight," I told my son and promised to check over the next couple of nights.

My heart swelled with love for this young man who still gets excited about sharing things with his mom, even though thirty years separates him from childhood. He is my oldest son and my most distant child, in miles. "You know what else?" he confided. "I found a special tree so I can send messages to you, too."

I'd told all the kids, years ago, that I had a particular tree under whose branches I would sit when I missed them. From there, I would send messages on the wind from my tree to every tree between us. When they felt a breeze on their faces, and heard the wind whispering through the branches, that was me, thinking of them.

Now here was Brendan, telling me he sent thoughts of me on the same ancient network, and I imagined the love of mother and child criss-crossing the globe, a comfort as old as love itself.

"Which one do you like?" asked my son Ken as we sorted through a bin of Beanie Babies in a gift shop. He was searching for an extra Christmas present for his youngest sister. My hand fell on a small, blue, spotted lizard. Its floppy little body draped over my fingers. Its forked red tongue and beady black eyes gave it a rather contemplative air. "Isn't it sweet?" I crooned.

Ken's girlfriend Trish agreed and we looked for the lizard on the price chart. "Wow!" I said, reluctantly putting the toy back in the bin. "That's a lot for a little bit of cloth and some beans."

"It would look neat on your computer, though," said Ken.

"I have a lizard on mine," added Trish.

I left them with their heads together over the bin, looking for just the right bean bag for Ken's gift.

The next day I opened the door to my computer cupboard. There, perched on the monitor, its front feet draped jauntily over the screen, was the little blue lizard, its beady eyes looking right at me.

I thought of Ken hiding it there and my heart swelled with love for this young man (and his sweetheart) who still thought his mom was special enough for surprises. I thought of our delight in pleasing one another, our happiness criss-crossing the globe, a joy as bright as love itself.

It is Wednesday evening and the phone rings right on cue. I know before I even answer that my eldest daughter will say, "Hi Mommy!" It's Jennifer night.

Once a week since she left for Florida several years ago, Jen and I speak, catching up on news or just chatting as if she were sitting next to me on the sofa instead of on her own, hundreds of miles away. Bob gets on the extension and the three of us laugh and carry on.

Over the long distance line we've shared sorrows and joys, disappointments and triumphs, nonsense and grave matters. My heart swells with love for this young woman who doesn't consider calling home every week a trivial matter. We are creating a web of devotion that criss-crosses the globe, a strength as great as love itself.

"Unwrap the small one first," Cassie instructs. There sits my youngest child, her face wreathed in smiles, her joy at surprising me written clearly in her eyes. I take the paper from the gift. It is a compact disk, and I know without unwrapping it that the larger present is a CD player. This child of mine, who knows how music can transport me, has given me a gift of more than clarity of sound. She has listened and watched and unerringly chosen something that will bring me joy.

Her pleasure in pleasing me means more than the gift itself, for in it I can see her heart. My own swells with love for this young woman whose happiness in sharing is a reflection of all that is good in the world. I think of our mutual delight criss-crossing the globe, an enchantment as contagious as love itself.

The Great Sled Race

When I was a child, snow often fell over the Berkshires shortly before Thanksgiving. It would catch first along the hedgerows and in the tall, dead grasses that lined the roads. Soon the lawns and fields were covered with a thin blanket of white and we children began to get excited. Toward the middle of December, when there was a good six inches of snow on the ground, enough to slide on, my brother Frank would take the sleds down from their storage place over the garage rafters, oil the runners well, and tie new knots in the frayed tow ropes. The twins and I, well wrapped in snowsuits and scarves and boots and mittens, would tromp up the hill behind him.

West's Hill was the best place to go sledding. The middle of the hill was quite steep and studded with large rocks that made marvelous jumps. We would plant our sleds at the crest of the hill, back up a bit to get a good running start, and belly flop on the sled, shrieking and yelling as we sailed down to the bottom.

There was a barbed wire fence that ran along the tree line at the bottom of the hill. Mr. West cut the fence near the middle, made a loop from a piece of the wire and dropped it over a cedar post, making a sort of gate. Most of the time we remembered to unhook the loop from the post and fold the fence wire back, leaving an opening wide enough for two sleds to go through side by side.

The January I was ten, we had a big race planned with the three children who lived on the hill. They argued that their three Flexible Flyers tied together could beat our new Christmas toboggan any day. We met at the top of the hill on a snowy Saturday afternoon to test their brag.

The wind blew the words out of our mouths as the twins and Frank and I argued about who would steer the toboggan and who would stand at the bottom of the hill to declare the winners. If the race was to be fair, only three of us could ride the toboggan down the hill. The fourth would have to act as the judge.

Frank dropped the toboggan onto the snow. He knelt on it and bounced a little to wedge it into the snow, making sure it would not take off without us. One of the twins was elected to referee the race. She trudged disconsolately down the hill and stood with her back to us, pouting. The other twin and I took our places on the toboggan.

The neighbor kids' three sleds were tied together. The oldest boy sat in front to steer, his brother sat on the middle sled and their little sister brought up the rear.

"We're ready!" Frank yelled downhill.

The referee looked up. She yelled something back.

"What?" Frank wanted to know.

The answer was indecipherable. She just stood, waving her arms and kicking the fence post.

"On your mark, get set, GO!" Frank hollered and with a running jump, landed solidly on the back of the toboggan. I heard our opponents give a whoop as the three sleds took off beside us.

We plowed headlong through the snow that blinded us as it blew back into our faces. The toboggan hit a boulder and veered off crazily, coming down with a great thump. The bottom of the hill was coming up fast. My little sister suddenly leaned back against me.

"Hemph," she shouted.

"What?" I yelled back.

"Fence!"

FENCE! The word was suddenly clear. In our excitement over the race, none of us had thought to move the barbed wire fence out of the way. Now a single strand of that wicked wire lay in wait across the path of our speeding sleds. The referee twin was frantically scrabbling with the loop but the deep snow had buried the lower strand of wire and she was too little to budge the gate.

"Bail out!" I screeched, grabbing my sister's shoulders, tipping us both to the left as hard as I could. Frank's feet, hooked around my waist, came along as we spilled into the snow. Right beside us I saw three bodies catapult off the sleds and plunge into the snow in a wild tangle of arms and legs. The toboggan and the sleds went on down the rest of the hill without us. We lay for a moment in stunned silence, then, "Tie!" hollered the referee and she stomped off home.

Freedom Ride

I was sitting on the back of my brother Frank's new Kawasaki Voyager, admiring the motorcycle's many fine points—the comfortable seats, the dual saddlebags, the stereo, the sparkling red fenders. I was perfectly safe. The bike, kickstand down, was parked in the driveway.

Frank handed me a helmet. I pulled it over my head. It fitted snugly on top, the wobble over the ears corrected by a tightening of the chin strap. I struggled a minute with the clasp. "How do you get out of this thing?" I asked.

We were waiting for his friend Mark to arrive and I didn't want to be caught sitting on the bike with my helmet on like an over-eager little kid impatient for the ride to start.

While we waited, Frank talked motorcycles. With one ear I listened to his description of the different types of bikes—sport, touring, off-road—and with the other I tried to gauge the threat of the thunder I could hear rumbling in the distance. I looked as he showed me the gauges and gears, the hand controls and the foot pedals while keeping one eye on the murderous thunderheads piling up in the western sky.

Just when I'd decided we probably wouldn't go after all, Frank said, "Well, let's do it," and climbed astride the bike. I perched behind him, looking for something to hang on to. I settled for the folds of his leather jacket as we pulled out of the driveway and onto the road in one smooth, fluid motion.

“Nothing to it,” I thought and then “Wa-ha!” I exclaimed as the road came rushing up on my left side. Too late I remembered Frank’s admonition not to lean against the curve.

“You all right?” he called back.

Upright again, my knees like jelly and my heart pounding, I said oh, sure, I was fine.

Without the protection of walls and a roof to block the onrush of air, it felt as though we were going at least the speed of sound but a glimpse of the speedometer said we were traveling at a mere 35 mph. We met Mark at the end of the street. He and Frank looked up at the sky where mud-colored clouds were gathering as if for battle.

“We can go south first,” Frank said. “It’s clearer that way.”

Mark smiled and shrugged. “I don’t have rain gear,” he said as if that settled it.

We rode off toward the Connecticut border, staying on the back roads. “The more curves and hills, the more fun,” Frank called back to me as we swooped and turned and swooped again. I gripped his jacket and tried not to think how close my feet were to the flying road, how exposed my jelly knees were as we sped along.

We followed the road to Salisbury and into Lakeville. A few drops of rain splattered against the windshield. We rumbled to a stop and Mark drew alongside. Frank looked at him. Mark shrugged. I was coming to recognize the body language.

We headed toward the west where the sky looked, if not clear, at least less ominous and rode out of the rain. And right toward the lightening.

The clouds, lurking behind the mountains, loomed suddenly large and black off to our right. Great, jagged streaks of orange and yellow jumped from cloud to cloud. We whooped each time we saw a flash. The air turned sharply colder and I shivered. I was more afraid of getting caught in the storm than I was of the proximity of my knees to the pavement. "Cold?" Frank called.

"Freezing," I yelled back, meaning "terrified," and saw with relief a pull-off area just ahead.

While I pulled on Frank's heavy yellow rain jacket, he and Mark discussed the best way to go. We were closer to Hillsdale, NY than Connecticut. The storm, racing just ahead of us, had spilled a torrent of rain along the highway though it was no longer falling where we stood. "Looks like we're following it," Frank said, eyeing the jumble of lightening-sparked clouds. "Let's just follow it home."

With the rain jacket breaking the wind I was warmer and with the storm moving away from us, I began to relax a bit. Dead ahead, the sky was clear and Jupiter was visible, shining like a beacon. Suddenly I realized that I'd been so worried about being struck by lightening, I'd forgotten to be afraid of hills and curves. My body had adjusted itself to the movement of the bike and somewhere along the way my knees had turned back into bone and tendon. I felt a swift surge of exhilaration. So close to home, I fervently wished we could go on forever, skimming the road with our feet, facing the wind and the winding road.

"Well," asked Frank, as we climbed off the bike and took off our helmets, "what do you think? Do you want to go again sometime?"

He was grinning. I was grinning. We had both tasted a wild freedom out there in the dark where the lightening flashed and we'd chased the clouds.

"It's more fun when the weather is nice," he said but I couldn't see how.

Lick, Spit and Polish

My sister Jeanne was dressing for dinner one evening. Her twin, Jackie, and I, both of us visiting Jeanne in Colorado, sat on the bed, watching. We were going to dinner, too, but we'd been ready for awhile. It doesn't take long to get dressed up when all you do is change your t-shirt.

Jeanne artfully tied a scarf around her neck, brushed her hair and gave it a spritz of hair spray, applied a bit of color to her eyelids and turned to look at us.

Jackie and I looked back at her, at each other, and then down at our own attire. We were clad alike in clean khaki slacks, t-shirts and sandals. So was Jeanne but somehow on her the khakis looked dressy, the t-shirt classy, the scarf a knockout touch.

Jackie heaved a big sigh. "It's always like this," she said gloomily. "I'm spit and she's polish."

"Where does that leave me?" I asked her, not really wanting to know.

"I guess you're lick," she joked. Neither of us laughed.

Jeanne has always had an undeniable something Jackie and I both lack; a flair for fashion, a keen eye for style, an enviable ability to make even her grungy gardening clothes look chic. Our mother had the same knack. In the most mundane of outfits she looked like a fashion plate. Ah, well.

As we waited for a table at the restaurant, Jackie and I watched the women coming through the door. "That's how I want to look," Jackie whispered as a tall, slim brunette swept in, her clingy knit dress showing off a perfect figure.

"You're not tall enough," I whispered back, envying the blonde dressed in a tank top and a pair of jeans. Svelte and sleek were beyond my dreams. All I really wanted to be was comfortable—stylishly.

A day or two later the three of us were dressing for a party. "It's casual," said Jeanne as she pulled on a flowered summer skirt. Jackie and I peered into our suitcases. Casual was all we had—shorts, slacks, jeans. Not a skirt in sight. I did have a sleeveless denim dress I'd packed at the last minute, just in case, but the day was cool and I didn't have a sweater. I pulled the dress over my head anyhow and rummaged around for a long sleeved shirt to keep the chill off. Jackie borrowed a scarf from Jeanne.

Jeanne looked lovely—neatly put together in a blue top that exactly matched the blue flowers in her skirt, and a delicate necklace that on me would have shouted, "Look at me, I'm all dressed up!"

Jackie and I looked hit or miss, mostly miss. My bare toes were chilly in sandals but it was cold feet or socks and sneakers, definitely déclassé. The shirt I'd thrown over my dress made me look like I'd dressed with a lick and a promise. I, too, heaved a big sigh.

I don't know why it should matter but it does. We discussed that, Jackie and I, as we watched Jeanne dress for work.

"What do I care what people think about my clothes?" I said as Jeanne tucked a color coordinated shirt into a pair of dress slacks, knowing as I said it that if it didn't matter, we wouldn't be having this conversation.

"Pretty is as pretty does," chanted Jackie, a thing we'd heard over and over from our mother.

We looked at one another. “So, Lick,” she asked, “what are you wearing today?”

“Spit,” I said, “does it matter?”

We both heaved a big sigh.

Missing Dad

My father has been gone for over twenty years yet I can conjure his face in a moment—high cheekbones, sharp nose, brown eyes that could look stern one moment and twinkly the next. I can hear the tenor of his voice and picture his tall, spare frame clad in khaki work clothes, a long-brimmed baseball cap pushed back from his forehead. He had jet black hair when he was a young man that grayed first at the temples and then all over. His skin seemed perpetually tanned, and his left forearm, which always rested on the rolled down car window when he made his mail rounds, was a shade darker than the rest.

He loved birds; he trained carrier pigeons during the war and raised chickens when it was over. A few of his close friends called him Jay Hawk, a name he secretly delighted in. He knew most birds by sight and many by their call. He hunted partridge and pheasant and quail. When I bemoaned the death of the beautiful ring-necked pheasant or the small, brown spotted quail, he would patiently explain the circle of life and death and life again as food for something else.

He could be wickedly funny, preferring slapstick humor to the puns and witticisms my mother was so good at. His favorite jokes were actual tales. He'd go on and on about a performing ant, for instance, before squashing the poor imaginary thing before our very eyes. "Which ant? This ant?" he'd ask gleefully grinding his thumb on the table as we kids groaned for being suckers. Or he'd intone, "It was a dark and stormy night..." and we knew we were in for ten minutes of, "What! Zinzindorf? Not Z I N -Z I N- D O R F?"

There were three occasions during the year when we made a fuss over my father—Christmas, his birthday, and Father's Day. He would never get up with the rest of us at the crack of dawn Christmas morning. That was left to my mother who sleepily followed us down the stairs and made hot chocolate while we tore the wrappings from our gifts. He claimed Christmas morning was the only time he did not *have* to get up and he enjoyed sleeping in until the hub-bub had somewhat subsided.

At nine-thirty or so, we were allowed to dash up the stairs and roust him out of bed. Then he would sit in his favorite green chair by the fireplace and we four kids would bring him his gifts. He would hold each one and shake it, turning it this way and that, trying to guess what was in it. Almost always, the box contained a tie or a sweater or another khaki work shirt but he made a game out of changing the ordinary into something special.

His birthday fell in April. My mother would bake his favorite chocolate cream pie and we children would pool our nickels and dimes to buy a box of thin mints, the only candy he really liked. He would slowly unwrap the box, knowing full well what was under the paper but pretending that the mints were the surprise of his life.

Father's Day brought more shirts and ties. One year my mother paid him back for all the Mother's Day irons and toasters and kitchen utensils by getting him a new lawn mower. Tied to the handle with a bright red bow was a box of thin mints.

There have been lots of holidays and birthday parties without my father since then but every Father's Day I still make a chocolate cream pie, open a box of thin mints and celebrate the memory of the man who taught me that life and death are the same thing and that life is more fun when you allow yourself to be surprised.

Remembering Mama

The old photograph shows her as just tall enough to nestle her head on my father's shoulder. "I'm built for comfort, not for speed," she joked once, but she could move quickly enough with hairbrush in hand if I was naughty.

Mama had a dancer's grace. I know she danced as a child—I became a ballerina in her cast-off pink satin toe slippers, and a tap dancing fool in her black patent leather tap shoes. "Like this," she would say, clicking her feet against the cement porch floor. "Brush, hop, tap, step, tap."

I repeated the words under my breath, *Brush, hop, tap, step, tap,* until my feet could do it without coaching. My mother would tap beside me, her sneakered feet slapping the beat beside my own.

Sometimes we'd stack her old polka records on the tall, silver spindle of the record player and hop madly across the living room, dodging the chairs, the footstool, the sofa, whirling and laughing till we were out of breath. When the music stopped, she went humming off into the kitchen to resume her ironing or start dinner.

Mama wore her hair in a sausage roll at the back of her head, even when it was cut short and the color had faded from flaxen to gray. I loved to watch her in the early morning as she stood in front of the mirror, her elbows bent, her hands reaching for the tiny hairs that escaped the rolled net that held her hair in place. On summer mornings she wore a morning coat—a thin, flowered, cotton dress with buttons all the way up the front. On winter mornings she bundled against the cold in a chenille robe, thick, and belted at the waist. Not until her housework was done did she dress for the day. She was in her fifties before she changed her skirts and blouses for shorts and slacks.

When I was a small child, where she was I also wanted to be. I would plead illness just to stay home from school and be with her. Her daily routine seldom varied. Up in the morning before the rest of us stirred, she had the kettle hot and breakfast ready when we stumbled, sleep-befuddled, into the kitchen. She drank two cups of morning coffee; one standing up,and the other while sitting at the table with my father as he drank his own.

Every weekday morning she packed four lunches—one for my brother, one for each of my sisters, and one for me in my much loved little green metal lunch box. Even when I stayed home sick, she'd pack my lunch in my lunch box. Then she'd fix a tray for herself, bring my lunch box to me in bed, and eat there in the sickroom with me.

She sang as she went about her daily tasks. "Singing makes it less like work," she said often, which may explain why I hum while dusting or washing the supper dishes. Often, too, she would stop whatever she was doing, seat herself before the piano in the living room and play for a half hour or so. I would scramble up on the bench beside her, watching her fingers fly along the keys. We would play together; two-part practice pieces by Bach or a madcap version of chopsticks, each of us playing faster and faster until one of us made a mistake. Then she would laugh and give my shoulders a one-armed hug.

When things went wrong or Mama was angry, she went out to the edge of the garden where an old apple tree stump served as a chopping block. She'd swing the ax, *thump!* against log after log until her anger had translated itself into a pile of firewood. If winter snow obliterated the chopping block, she'd coax us children to go skating or sledding with her. Before long she'd be having so much fun that her bad humor would vanish in the thin, cold air.

I remember all these things about her as I put the photo album back on the shelf. I do a little *brush, hop, step* as I head for the kitchen to make dinner.